Forest Biotechnology and its Responsible Use

A publication by the Institute of Forest Biotechnology

Forest Biotechnology and its Responsible Use

A publication by the Institute of Forest Biotechnology

Raleigh, North Carolina | USA | Published by the Institute of Forest Biotechnology

Contents

Preface

This document discusses the basic concepts of forest biotechnology, how biotech trees are managed, and why the Institute of Forest Biotechnology created a set of stewardship principles for the responsible use of these trees. We hope that you find it informative and thought provoking.

The Institute of Forest Biotechnology is the only organization to address the sustainability of forest biotechnology on a global scale. With the help of our Partners and Sponsors, we bring diverse stakeholders together to address societal, environmental, and economic aspects of forest biotechnology.

Our forests are under pressure from global trade, population growth, invasive threats, and increased demand on natural resources. The IFB supports responsible uses of forest biotechnology that benefit society and the environment through science, dialogue, and stewardship. Together, with the Forest Biotechnology Partnership, the IFB is the most comprehensive information source on forest biotechnology – anywhere.

I would like to thank the Responsible Use Initiative's Implementation Committee, the U.S. Forest Service, and our Sponsors for providing the technical and financial resources to make this work possible.

For additional information on the Responsible Use: Biotech Tree Principles, please visit www.responsibleuse.org. This website has the latest version of the Principles, and useful information about biotech tree stewardship.

Thank you for your interest in protecting the future of forests. Please contact us if you have any questions about forest biotechnology in general, or if you would like to participate in the Responsible Use: Biotech Tree Principles.

Thank you,

Adam Costanza - President   Susan McCord - Executive Director

Overview

Biotechnology is being used as a tool to grow trees with special characteristics. When used responsibly, society and the environment can benefit from advanced tree breeding technologies. The next decade will be a time of rapid expansion for biotech trees throughout the world in an attempt to meet global demand for forest products, biofuels, to restore threatened species, and to protect future forests from invasive pests and climate change. The world will benefit from a mechanism to determine which uses of these technologies are most likely to bring benefit. The Institute of Forest Biotechnology (IFB), along with stakeholders from around the world, have developed the Responsible Use: Biotech Tree Principles.

This document focuses on the science, dialogue, and stewardship of forest biotechnology to help foster clarity and build a foundation for ongoing discussions. There are a lot of topics discussed and explained in plain terms. In order to keep this document concise and easy to read some terms are used without definition or lengthy discussion. For more in-depth information we recommend the following resources:

- www.forestbiotech.org - IFB's main website with extensive resources on forest biotechnology

- www.responsibleuse.org - Responsible Use: Biotech Tree Principles and biotech tree stewardship resources

- www.treegenes.org - Directory, glossary, and background on tree gene related work

The Science of Forest Biotechnology

"The term biotechnology[1] came into common usage in the 1980s, and its several definitions continue to change. Broadly defined, it is anything that combines biology and technology. Biotechnology has become more controversial as the level of the technology has increased." An excerpt from "Genetically Modified Forests from Stone Age to Modern Biotechnology" (Burdon et. al. 2006). This book is the most comprehensive review of forest biotechnology to date. Within the context of forest biotechnology we can refer to 'modern' biotechnology as that which postdates the discovery of the structure of deoxyribonucleic acid, or DNA. Modern biotechnology is centered on the analysis and manipulation of DNA and the insertion of DNA fragments into organisms. Biotechnology has advanced to where we can research biological systems at the DNA level, and even sequence entire genomes. This feat was accomplished through years of research in plant and agricultural life sciences. While we can sequence a human's genome today, it is not as simple for some trees. For example, a representative conifer (pine tree) genome is many times larger than a human's (Keim, 2008); the Loblolly pine's genome is seven times larger. Such a feat has yet to be accomplished in whole, but the Pine Genome Initiative[2] is working towards this goal to improve future biofuel production from pine trees, advanced wood products, enhanced carbon sequestration, improve forest health, and gain insight into plant evolution.

As we consider the spectrum of techniques that can be considered forest biotechnologies, it is useful to split out two general groups: breeding technologies, and genetic engineering technologies.

Breeding Technologies

Breeding technologies involve different applications of biotechnology. The various breeding technologies described in this section are ordered from least to most advanced in terms of their reliance on advanced techniques for success. Be aware that one or more of these breeding technologies are used to produce genetically engineered[3] trees.

Conventional Breeding

Purposeful breeding of trees involves producing hybrids, progeny of parents from different genetic populations, to create a desired phenotype. An example of a hybrid would be a cross between a tree that has exceptional growth and one that shows resistance to a fungus that attacks that particular species. This is an imprecise method of producing a tree with a specific characteristic because there is no control over additional genetic material being incorporated along with the desired phenotype.

---

1   "Biotechnology: Any technological application that uses biological systems, living organisms, or derivatives thereof, to make or modify products or processes for specific use." (Convention on Biological Diversity, 1995)

2   Pine Genome Initiative: www.pinegenomeinitiative.org

3   "Genetically Engineered and Genetic Engineering: The use of recombinant DNA and asexual gene transfer methods to alter the structure or expression of specific genes and traits." (Food and Agriculture Organization of the United Nations, 2004)

Asexual Propagation

Also referred to as clonal or vegetative propagation, is in contrast to sexual reproduction. Sexual reproduction requires fertilization via pollen and is how most plants reproduce. Asexual propagation produces genetically identical trees; a process which does occur naturally by some tree species. For example, breaking off a branch of a willow tree and sticking it in the ground will produce roots from the branch. Over time a new willow that is genetically identical to the original tree will grow. Foresters have used this technology to their advantage as a breeding tool for centuries. Propagating a tree by grafting or with rooted cuttings allows the production of individuals that are genetically identical to the original and are therefore members of a clonal family. Tissue culture has become the preferred method of clonal production for some species of plants including several species of forest trees. This technology is also useful for moving clones across quarantine barriers since the cultures are free of insects and diseases and can easily be grown into mature plants.

Organogenesis

A type of asexual propagation that literally means 'organ genesis'. It is accomplished by growing a mass of cells in tissue culture that have the ability to produce shoots and grow into full trees.

Somatic Embryogenesis

Cells in a tree other than the pollen and egg cells are called somatic cells. This technology-intensive technique is used to rapidly proliferate plant tissue via asexual propagation that mimics steps of the normal embryo development process. Single cells in tissue cultures from shoots, buds, roots, or leaves are induced to form complete embryos that can form individuals identical to the original plant where the cells were taken. Large numbers of a given clone can be produced in a short time with this technology.

Marker Assisted Selection

When researchers have enough information about a tree's genome, they can use molecular markers to locate specific genes in potential seedlings, at an embryonic stage. These molecular markers are naturally in the tree's DNA. These markers are easier to identify than specific genes. By serving as constant landmarks in the genome, markers give breeders the advantages of speed and accuracy in their breeding program.[4] With this technology they are able to determine if a tree, at the embryonic stage, has the trait of interest instead of waiting years for the tree to grow and mature.

---

4   Marker Assisted Selection from the Food and Agriculture Organization of the United Nations: http://www.fao.org/docrep/010/a1120e/a1120e00.HTM

Genetic Engineering Technologies

Genetic engineering is an advanced science in which DNA sequences encoding for very specific and desired traits are introduced into the plant genome. We must be precise in our language and terms because as discussed above, forest biotechnology includes a wide range of technologies that are used to improve trees.

Certain types of forest biotechnology require the most immediate attention for stewardship principles to guide their use, such as through the IFB's Responsible Use: Biotech Tree Principles.[5] Clonal propagation of trees is not necessarily the result of biotechnology. Some types of trees in the forest, such as willows, can asexually propagate through shoots or branches without any human intervention. Conventional breeding techniques that include various crossing and selection methods can produce trees that can be asexually propagated. Both breeding and asexual propagation have been used for over 3,000 years and are generally regarded as familiar and safe. Society should focus its efforts on developing principles for the truly advanced forest biotechnologies rather than on trees that were originally developed through conventional breeding techniques.

The Institute of Forest Biotechnology defines 'biotech trees' as: "trees developed through genetic engineering or which contain discretely engineered DNA, and their offspring.[6]" This definition is intentionally inclusive of both the process, being developed through genetic engineering, and the resulting tree that contains engineered DNA. The IFB considers the offspring of a biotech tree to also be a biotech tree unless it is known that the tree does not contain genetically engineered DNA. For example, if a biotech tree is crossbred with a non-biotech tree then the resulting offspring may or may not contain the engineered genes present in the biotech tree parent. As long as it is unknown if the offspring contains engineered DNA, the IFB considers it to be a biotech tree. Given this definition, biotech trees encompass a spectrum of advanced technologies that introduce genes into trees via genetic engineering. The IFB uses the term biotech tree because there is a need for more clarity in this field where there are a lot of similar and confusing terms. For example, a generic term commonly used for transferred genes is 'transgenes.'[7] Transgenes are simply genes moved from one organism to another. However, the term 'transgenics' or 'transgenesis' refers only to the process of introducing genes from an unrelated organism that could not naturally breed. There are a number of technologies that can produce a biotech tree. These technologies and terms are outlined here to help foster the science, dialogue, and stewardship of forest biotechnology.

---

5   Responsible Use: Biotech Tree Principles is an initiative of the Institute of Forest Biotechnology to create principles for the stewardship of biotech trees. The principles were developed through a multi-stakeholder, transparent process. More information is available at www.responsibleuse.org

6   www.treegenes.org/glossary.html#biotech-tree

7   "Transgenes: a gene or gene construct that has been transferred into an organism such as a plant, using genetic engineering techniques, including transformation techniques, i.e. the process of inserting transgenes into the genetic material (DNA) of an organism." (Convention on Biological Diversity, 2006)

Altered native gene function

This technique does not insert a gene, so in some instances it may not be considered transgenic,[8] but the IFB considers them biotech trees nonetheless. Instead, a regulatory section of DNA, often called a promoter, is inserted into the gene of interest. This regulatory section can cause an increase or decrease (upregulation or downregulation) of the expression of native genes. By altering how a gene is expressed, scientists can adjust phenotypic aspects of the tree. This technique can be used to increase the growth of trees, control flowering, and modify wood composition. Often these types of trees mimic those already in natural tree populations.

Cisgenesis - Insertion of one or more novel genes from plants that could naturally cross

Cisgenesis involves taking genes found in wild relatives of the tree in which they are inserted. Since the genes are from trees that could naturally interbreed, conventional breeding techniques may achieve the same results if enough time and effort were expended. Because of the long generation times of trees it may take decades for a conventional breeding program to produce the same results as a cisgenesis approach. This technique can be used to produce trees that have desirable attributes from other trees. For example, transferring the genes that confer resistance to the Chestnut blight from the Chinese chestnut and inserting them into an American chestnut would use this technique.

Insertion of one or more novel genes from plants that could not naturally cross

This form of transgenesis involves taking genes that would not typically be found in the normal gene pool for the tree in which they are inserted. The genes inserted are from other plant or tree species to produce novel trees that would not normally occur through natural selection on a human timescale. However, since the genes do come from the same kingdom it is within the realm of possibilities that given enough time, those genes could naturally become part of the tree's genome. This technique is being used to produce trees that are resistant to disease, drought, pesticides, and cold temperatures.

Insertion of one or more novel genes from non-plants

This form of transgenesis involves taking genes from organisms in a different kingdom. This application produces novel trees that would not normally occur through natural selection because the genes come from organisms other than plants. Trees engineered to absorb pollutants from the environment rely on these types of transgenes.

---

8   Regulating a native gene in this manner still requires that DNA is inserted and therefore it can be regulated as a 'transgenic technology' even though it is not an actual gene being inserted. For example, in the United States, promoter insertions would be regulated as transgenic.

Transgenes providing new industrial applications

These introduced genes can use any type of transgenesis mentioned above. Usually the resulting tree is used as a vector to produce material for industrial applications and would not naturally occur. For example, it may be possible to engineer a tree to produce pharmaceuticals, or chemical feedstock intermediaries. This technology is on the very cutting edge of forest biotechnology today.

Not genetic engineering - Mutagenesis

Mutagenesis[9] is discussed here only to clarify why it is not considered a targeted genetic engineering technology. Some new varieties of trees, such as new fruit tree varieties, have been generated from the selection of natural or induced mutations that can alter color, smoothness, shape, and seed characteristics. Mutated branches can be asexually propagated to preserve and maintain the new characteristic resulting from the mutation. Although these trees do contain genes that are modified, the process of non-targeted human induced mutation are not considered genetic engineering here and do not fall within the IFB's definition of a biotech tree in part because many plant mutations are deleterious to the plant and only a few are able to become desirable new varieties. This is mainly because mutagenesis has multiple large impacts on the plant genome that cause a pleiotrophic, or chain reaction of effects in the genetic structure of the tree, that are usually both positive and negative. Wide ranging and random effects on the genetic structure of any organism are rarely beneficial to its survival. Today this technology is unlikely to be used in place of genetic engineering in trees where generation times severely restrict looking for a needle (the phenotype you want) in the haystack (all the cell lines you would have to generate and test). In short, mutagenesis is not a form of targeted genetic engineering. It is a random approach that alters a suite of genes in unknown ways.

---

9   Human induced mutagenesis changes a genetic structure through the use of mutagens such as ultraviolet light, radiation, or chemicals. It is fair to say that inducing mutations, whether novel or not, rapidly alters the genetic composition of an organism. However, this is not biotechnology that uses living systems to produce the mutated plant, and it is not genetic engineering that uses targeted DNA sequences encoding for very specific and desired traits.

The Dialogue of Forest Biotechnology

While it is impossible to rank-order every possible benefit and risk associated with a technology, we can reasonably expect society to make a distinction for trees that differ most from their native relatives based on the assumption that unfamiliar traits may pose more risk to humans or the environment. It will be society's concerns which will set the boundaries in the commercialization and stewardship of biotech trees.

In many ways, the source of society's concern involves speed. It has been argued, and will continue to be in some circles, that what humans are doing with genetic engineering is a faster version of the natural processes occurring all the time. Nothing is static. Species evolve over time. The 'natural' state of the earth's ecosystem is dynamic and always changing. For example, human beings are generally thankful for the mutations that have made us distinct from our ancestors. Together mutations and natural selection have created the myriad species on earth today, including the vast variety of trees in our forests. What we are able to do with biotechnology is twofold. We are able to speed up natural evolutionary processes that we can reasonably expect would happen over many millennia. An example of 'accelerated evolution' could be creating a tree that is resistant to a particular virus. We are also able to extend the abilities of trees in ways that we can reasonably expect would never happen naturally. An example of 'novel evolution' could be creating a tree that produces pharmaceutical products. Many of our drugs today are derived from tree products, such as tamoxifen (Geffen et. al. 2001), a drug originating from trees and used to fight breast cancer. In the future there may be trees that are engineered to produce products that are not naturally tree-based.

Biotech trees will change future forests. There are a number of ways that society can address the appropriate use of this technology: through laws and regulations, certification programs, and industrial pledges. Since society places a high value on the earth's forests and the ecosystem services they provide, we believe that social interests and the ecological need for science-based stewardship warrants more comprehensive options to address the use of forest biotechnology, and provide assurance that biotech trees can be used responsibly. To fill this gap in performance the Institute of Forest Biotechnology created the Responsible Use: Biotech Tree Principles to foster stewardship of these trees. This initiative focuses only on biotech trees; those depicted by the yellow and orange colors in figure 1. In very broad terms, increasing technical advances in biotechnology can potentially lead to greater deviations from native phenotypes. In other words, the more gene function is modified in a particular tree, the more likely that tree will have different characteristics from its native relatives. As figure 1 shows, there is a higher potential to produce trees with characteristics that are different or altered from native phenotypes as the amount of technology applied increases.

However, more technology does not necessarily mean more genes are being modified. Many of the advances in biotechnology are to make gene modification more precise and thereby limit the extent of potential risks. This is an important distinction because a common belief is that applying increasingly more modern biotechnology techniques equates to more significant changes in an organism's DNA. In reality, genetic engineering makes much more precise changes at the DNA level than traditional breeding. Even a considerable amount of genetic engineering would pale in comparison to the large DNA alterations that traditional breeding makes. It is the specific, targeted, and unique changes through modern biotechnology that make phenotype changes possible that differ from native ones. The graphic below is showing the potential changes from native phenotypes when more technology is applied.

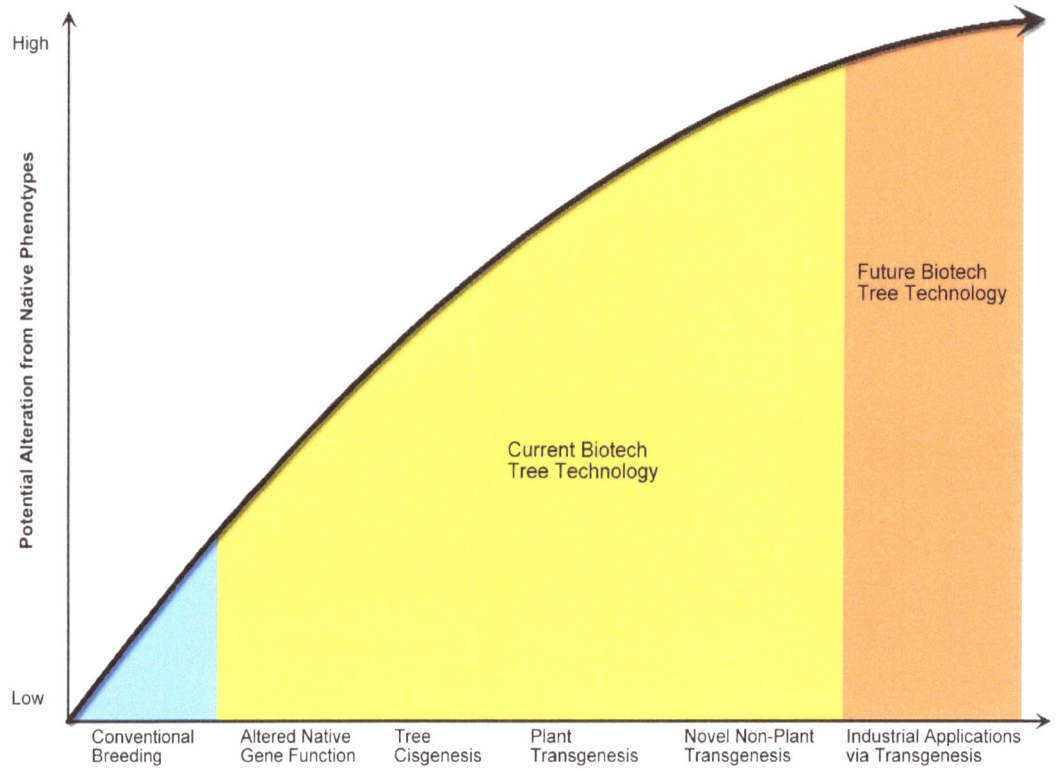

Figure 1

Stewardship of Biotech Trees

T he Responsible Use: Biotech Tree Principles are a set of practices that people can use to guide stewardship of biotech trees. There are various types of control imparted on biotech trees by regulatory systems around the world that are described generally in the next section. Rather than duplicate these systems, the Responsible Use Principles create a high level of stewardship that is performance based.

How biotech trees are controlled

There are four main levels of containment for biotech trees, with varying types of release control throughout the world. Control requirements typically start in the lab under highly controlled environments like greenhouses and then continue to outdoor test plots with a high level of oversight. Finally, some biotech trees are released to the open environment without the requirement for any control. The reason these four distinct levels exist is based on both the need to study the phenotypic aspects of growing trees, and the potential for biotech trees to interact with the natural environment. Each level has its own unique aspects that are managed in different ways.

Level 1 – Confined to the lab or greenhouse

By confining early results of genetically engineered trees to highly controlled environments, there is almost zero chance that any genetic introgression[10] will occur with trees in the natural environment. This level of control allows scientists to conduct research towards improving and understanding the biology of trees without the risk of affecting living systems outside of the lab or greenhouse. The downside is that labs are much smaller than outdoor plots so space often limits how many trees can be tested, or how long they can be allowed to grow since most trees will grow many meters high.

Level 2 – Field trials with oversight

The outside environment is highly dynamic and can never be perfectly replicated in the lab. For this reason many biotech trees are moved into fields where they are exposed to natural conditions. The spatial and temporal distribution of biotech trees is tightly controlled in field tests in the United States, while other countries may require different levels of specific oversight. This level of control allows researchers to grow more trees, usually to an older age, in real-world conditions, to test the performance of the trees and the efficacy of the inserted DNA sequences. In most cases this level requires that each biotech tree planted in the field be managed and tracked according to a specific plan established by the regulating authorities.

---

10   Genetic introgression: The infiltration of the genes of one species into the gene pool of another.

<u>Level 3 – Released for planting with monitoring requirements</u>

While there are a number of specific details based on the regulatory system and the biotech tree itself, this level of control is less restrictive on the spatial and temporal range of the tree. It is possible that a regulatory agency could allow release on the condition that certain aspects of trees and plantings are monitored for potential environmental risks.

<u>Level 4 – Released for planting without monitoring requirements</u>

At this level, a biotech tree is treated no differently than any other planted tree, and there is no requirement for monitoring or regulatory oversight.[11] To date, there are only two biotech trees at this level in the world today: virus resistant papaya (United States) (Zakour, 1998), and insect tolerant poplar (China). Another tree, the plum pox virus resistant plum (United States) is awaiting final approval for release at this level at the time of printing (Scorza, 2010).

There are orders of magnitude in the number of trees at each level. With thousands of highly confined biotech tree specimens and several hundred field tests around the world there are still only two biotech trees that do not require regulatory oversight and can be planted like any other tree. This relatively small number reflects in part the stringency by which biotech trees are tested and move through some regulatory approval processes. However, some of the most powerful forces slowing biotech tree progression between levels of control are the high costs of developing and testing biotech trees. In addition, public perceptions of biotech trees need to be more thoroughly investigated. The IFB in collaboration with its Forest Biotechnology Parters is working to better understand public perceptions regarding biotech trees.

Status of biotech trees in 2010

As of the date of this publication, there were two species of biotech trees in the environment that are either deregulated or not regulated, meaning that there is no governing body looking specifically at these trees and how they interact with the environment and society. Information regarding the number, type, and governing body of non-regulated biotech trees is available at the IFB's website.[12]

<u>Deregulated – Papaya, U.S. 1997</u>

The three U.S. agencies that have regulatory oversight (U.S. Environmental Protection Agency, U.S. Food and Drug Administration, U.S. Department of Agriculture's Animal and Plant Health Inspection Service) gave the papaya a non-regulated status for use in Hawaii to combat a virus. This made the tree available to any farmer interested in planting it. Because papaya is a food crop it had to be reviewed by all three regulating agencies. An additional variety of Papaya,

---

11   U.S. Department of Agriculture's Animal and Plant Health Inspection Service (APHIS) has the authority to repeal an unregulated status on any plant if it subsequently becomes detrimental to society.

12   Updated list of biotech trees is available at http://www.forestbiotech.org/biotechtrees

X17-2, was deregulated in the U.S. on September 2, 2009.[13] This variety is also engineered to withstand the papaya ringspot virus.

Not regulated – Poplar, China 2002

China commercially released 1.4 million genetically modified Poplar (*Populus*) trees in an area of 300 – 500 hectares (Food and Agriculture Organization of the United Nations, 2004). As opposed to the Papaya, which is considered a food crop, this event marked the first biotech forest tree ever released into the environment. These trees were modified with the Bt gene that produces a protein toxic to insect pests. This gene is produced by a bacterium in the soil called *Bacillus thuringensis* and is used in a number of biotech crops including corn, cotton, and soybeans. Today there are spurious reports that these trees are now readily available to tree farmers to plant for various purposes.

Pending deregulation – Plum, U.S. 2009

In 2007, the U.S. Department of Agriculture deregulated the C5 "HoneySweet" Plum tree engineered to be resistant to the plum pox virus (Animal and Plant Health Inspection Service, 2007). Like the Papaya, this tree is also a food crop and was subject to similar requirements. As of printing, this tree is awaiting approval from the U.S. Environmental Protection Agency. The HoneySweet Plum was issued a letter of "no further questions" by the U.S. Food and Drug Administration in January 2009, which is effectively an approval to use. U.S. Environmental Protection Agency has conditionally registered this tree, and is expected to fully register it in May, 2011.

Of these tree species the million plus Poplars released in China is the most interesting test case because these forest trees can cross with native relatives. This can change the genetic make-up of the natural forests; for better or worse, or there may be no significant change at all. There is of course no going back – modified genes will be in future forests. This fact has raised the question of whether society should put principles in place for users of biotech trees to follow.

Petitioned for deregulation[14] – Apple, U.S. 2010

The "Non Browning Apple" from Okanagan Specialty Fruits Inc. in Canada has petitioned the U.S. Department of Agriculture's Animal and Plant Health Inspection Service for deregulation of their apple tree that produces fruit that do not brown when cut. More information is available at the company's website: www.okspecialtyfruits.com.

Petitioned for deregulation[15] – Eucalyptus, U.S. 2010

The "Freeze Tolerant Eucalyptus" from ArborGen Inc. in the U.S. has petitioned the U.S.

---

13   http://www.aphis.usda.gov/newsroom/content/2009/09/printable/gepapayav_brs.pdf
14   http://www.aphis.usda.gov/biotechnology/not_reg.html

Department of Agriculture's Animal and Plant Health Inspection Service for deregulation of their tropical Eucalyptus tree that has the ability to withstand freezing temperatures. More information is available at the company's website: www.arborgen.com.

Risks and benefits of using biotech trees

Biotechnology is a powerful tool. Humans have had the ability to change living organisms for thousands of years and have had to reckon with consequences ever since they began breeding and translocating plants. Forest biotechnologies that modify genetic operations are no exception.

However, unlike the implementation of some technologies that mankind has embraced with disregard for social and environmental implications, such as the introduction of the automobile, many experts believe that the use of biotech trees can take a radically different course. To date, biotech tree work has focused on either environmental or economic benefits. This fact is in large part because of the technical difficulties that have to be overcome to genetically modify trees. The difficulties make it very costly and time consuming to make even minor adjustments in trees. The major institutions around the world doing this work are thoughtful in their approaches with many checks and balances along their research chain that ensure resulting biotech trees are useful to society. In the end there has to be a significant ecological, social, or economic return, or the research is discontinued for more relevant work. The rigor of this process does not eliminate all risks or unintended consequences associated with biotech trees, but it has created broad categories of risks and benefits that are generally accepted among forest biotechnologists as important to consider during the research and development stage.

Potential benefits of using biotech trees

The following are some beneficial uses for biotech trees identified by the Forest Biotechnology Partnership.[15]

Enhance bio-based products

Trees can be engineered to adjust the characteristics of its wood to suit specific needs. The global demand for biofuels has put an unprecedented amount of resources into developing trees that can readily be converted into liquid fuels. For example, adjusting the ratio of lignin and cellulose in wood can create a feedstock that is significantly easier to convert into a liquid fuel (United States Department of Agriculture, 2007). Whether it is ethanol, butanol, synthetic gasoline, or even biodiesel, the interest in advanced forest fuels will forever change how society considers purpose-grown trees. However, it is not just the race to produce cellulosic biofuels that is putting biotech trees on the fast track to commercial production. Conventional forest

---

15   The Forest Biotechnology Partnership is made of expert individuals and organizations working with the Institute of Forest Biotechnology to promote science, dialogue, and stewardship in the field of forest biotechnology. More information is available at: http://partners.forestbiotech.org

products can benefit from biotechnology as well. Developing trees that grow faster, denser, and straighter is desirable to lumber companies. Making more cellulose and less lignin means more pulp and paper can be produced from a single tree. The list of commercial benefits from biotech trees is long and must include the ancillary benefits of providing renewable products that society uses.

Combat invasive threats

Engineering trees so they are more resilient to changing climates and are better able to defend against foreign pests is critical to keep our forests healthy. Earth's climate is changing at the same time invasive pests are able to travel from one side of the earth to the other in less than 24 hours in cargo holds. Forests evolved over millennia of relatively stable climates and large spatial separations. Demanding that native trees combat the onslaughts of threats they encounter today without human intervention will result in significant tree deaths and ecosystem change around the globe. This is already being seen in British Columbia, Canada, where, the mountain pine bark beetle is estimated to claim 80 percent of the mature Lodgepole pine by 2013 (Natural Resources Canada Program Description, 2007).

The Forest Health Initiative[16] is a collaborative effort to advance the understanding and role of biotechnology in addressing forest health challenges. The initiative is initially focusing on the American chestnut as a test species. The initiative uses a holistic approach to address emerging forest health threats by assessing not just the science but the societal and policy issues concurrently. More information about the Forest Health Initiative is available online at www.foresthealthinitiative.org.

Maximize forest productivity

Growing more material on less land more quickly has potential benefits to the forest products industry, the new liquid biofuels industry, and as a method to sequester carbon more quickly. A significant amount of research has gone into biotech trees that make more fiber on less land. These trees have the potential to reduce demand on native forests and make it easier to use trees as a fast rotation crop in intensely managed tree plantations. Dense wood crops may also play an important role in international incentive-based programs, such as the United Nations Collaborative Programme on Reducing Emissions from Deforestation and Forest Degradation in Developing Countries (UN-REDD). UN-REDD is designed to avoid deforestation, slow forest degradation, and reduce the economic incentives that promote illegal logging (UN-REDD Framework Document, 2008).

---

16   The Forest Health Initiative is not an initiative of the IFB. However, the IFB is involved in the initiative as project secretariat and policy lead.

Replenish resources

In many communities around the world, the forests that people have historically relied on are in decline. The reasons are multifaceted but the fact remains that people are struggling, forests are disappearing, and these two situations reinforce one another through complex negative feedbacks. The solution is not as easy as producing biotech trees to reclaim degraded land, but advanced societies interested in improving these situations should not overlook forest biotechnology tools when novel solutions are needed.

Potential risks of using biotech trees

A recent symposium of international experts highlighted the following areas of risk that should be addressed when biotech trees are used in the open environment (Institute of Forest Biotechnology, 2008). Summaries of the symposium findings are given below.[17]

Gene flow and introgression

Gene flow refers to the introduction of a transgene into a native population of the same tree species. Before gene flow can occur there must be seed or pollen dispersal, establishment of the propagule, and survival of the plant to sexual maturity. In some situations gene flow is the desired outcome of releasing biotech trees into the environment. Forest health projects to protect threatened trees species from invasive threats plan on gene flow from genetically modified trees to native ones and thereby confer resistance to the native population. In some instances gene flow will have to be managed appropriately if the intent is to protect native forests from genetically modified tree genes. Introgression is the infiltration of a transgene from one species into the gene pool of another. This can occur through repeated backcrossing of a hybrid with one of its parents. This situation has similar ramifications to gene flow, but it adds a confounding factor of moving a transgene from one species to another.

Exceptional fitness

While one of the main goals is to produce trees that are better able to thrive in the environment, there are concerns that these more biologically fit trees may outcompete native species. This characteristic is commonly referred to as weediness. If a biotech tree is exceptionally fit when compared to its native relatives, there is a chance that it will be so successful in the forest that it keeps other trees from growing as they naturally would. Again, there are instances when exceptional fitness is the desired outcome as well as situations where it can cause ecological problems.

---

17   A PDF of the book, "Genetically Engineered Forest Trees: Identifying Priorities for Ecological Risk Assessment" is available free at the IFB's website:
http://www.forestbiotech.org/pdf/GE_Trees_Ecorisk_online_v1.pdf

Effects on non-target species

Possible effects to non-target species might include harm to beneficial soil organisms, insects, birds, or other plants. For example the Bt gene that is one of the more common genetic modifications in plants produces a protein toxic to some insect pests. Using the Bt Poplars planted in China as an example, we can envision a situation where the Bt gene might inadvertently harm a non-target species of insect. This specific situation is extremely unlikely in appropriately constructed and well-managed trees because of the rigorous research that has gone into the Bt gene, but it is nonetheless a concern. Situations where thorough research on the expressed gene has not been completed could potentially harm organisms that interact with the biotech tree.

Biodiversity effects

These concerns are broader and involve interrelations among forest species that affect whole ecosystems. One such concern focuses on the idea that stands of sterile GE trees would not support a diverse population of species in the larger forest ecosystem. Another concern is that the target function of the biotech tree will have unintended ecosystem consequences. A situation could be envisioned where a biotech tree is exceptionally fit, contains a transgene that inadvertently affects non-target species, and there is gene flow to native trees. This situation could have negative biodiversity effects for the native forest.

These concerns highlight scientific questions that need to be addressed through ecological research. In addition, there are social concerns that are valid even if all of the scientific concerns are addressed. The proliferation of sustainable forestry certification schemes shows that the public places a high value on responsible management of natural resources and the well being of forest ecosystems. By recognizing the ecosystem-changing potential that irresponsible biotech tree use may cause, we are in the enviable position today of being able to construct and implement stewardship practices before biotech trees reach the market en masse. That is why the IFB has created the Responsible Use: Biotech Tree Principles, an initiative to develop these critical practices in a highly transparent and multistakeholder driven process, available at www.responsibleuse.org.

References

The Animal and Plant Health Inspection Service (APHIS), "Approval of USDA-ARS Request (04-264-01P) Seeking a Determination of Non-regulated Status for C5 Plum Resistant to Plum Pox Virus" United States Department of Agriculture (USDA). June 27, 2007. www.aphis.usda.gov/brs/aphisdocs2/04_26401p_com.pdf

Burdon, Rowland D., William Libby and Robert Kellison. "Genetically Modified Forests from Stone Age to Modern Biotechnology" Forest History Society Issues Series, Durham NC, 2006.

Convention on Biological Diversity. June 1995. Online: http://www.cbd.int/convention/convention.shtml

Food and Agriculture Organization of the United Nations. Preliminary review of biotechnology in forestry, December 2004. Summary available online at: http://www.fao.org/docrep/008/ae574e/AE574E03.htm

Forest Genetic Resources Working Paper. "Preliminary review of biotechnology in forestry, including genetic modification." Food and Agriculture Organization of the United Nations – Forestry Department. 2004. ftp://ftp.fao.org/docrep/fao/008/ae574e/ae574e00.pdf

Geffen, David B. MD and Sophia Man MD. "New Drugs for the Treatment of Cancer, 1990-2001", Department of Oncology, Soroka University Medical Center and Faculty of Health Sciences, Ben-Gurion University of the Negev, Beer Sheva, Israel.

"Genetically Engineered Forest Trees: Identifying Priorities for Ecological Risk Assessment" The Institute of Forest Biotechnology, September, 2008. http://forestbiotech.org/pdf/GE_Trees_Ecorisk_online_v1.pdf

Institute of Forest Biotechnology. www.forestbiotech.org

Keim, Brandon. "The Cheapest Genome Sequence Ever: For Real?" Wired Science Online. October 6, 2008. http://blog.wired.com/wiredscience/2008/10/the-cheapest-ge.html

Natural Resources Canada. "Forest. Forward. Moving Beyond the Pine Beetle" Program Description. http://mpb.cfs.nrcan.gc.ca/index_e.html

Pine Genome Initiative, 2007. http://www.pinegenomeinitiative.org

Scorza, Ralph. Personal communication regarding the regulatory status of the Honeysweet Plum tree March 6, 2009. Information available online: "Honeysweet Plun Trees: A Transgenic Answer to the Plum Pox Problem", http://www.ars.usda.gov/is/br/plumpox/

Submission to the Convention on Biological Diversity on Advice of the Report of the Ad Hoc Technical Expert Group on Genetic use Restriction Technologies. Convention on Biological Diversity, 2006. http://www.cbd.int/doc/meetings/tk/wg8j-04/information/wg8j-04-inf-17-en.pdf

United States Department of Agriculture. "Trees, A Poplar Source for Biofuels", Cooperative State Research, Education, and Extension Service. October, 2007. http://www.csrees.usda.gov/newsroom/research/2007/poplars.html

UN-REDD Framework Document. "UN Collaborative Programme on Reducing Emissions from Deforestation and Forest Degradation in Developing Countries", Environmental Defense Fund, 2008. http://www.edf.org/documents/7975_REDDandCarbonMarketAnalysisReport_EDF_0508.pdf

Zakour, John and Linda McCandless. "First Genetically Engineered Papaya Released to Growers in Hawaii" April 18, 1998. Available online at: http://www.nysaes.cornell.edu/pubs/press/1998/papayarelease.html

Notes

www.ingramcontent.com/pod-product-compliance
Lightning Source LLC
Chambersburg PA
CBHW060814290526

45792CB00005BA/1652